God Made the Dinosaurs Long Ago

Carol Pirog

Anchor Book Press • Palatine

God Made the Dinosaurs Long Ago

Imprint: Anchor Book Press
440 W Colfax Street, Unit 1132, Palatine, IL 60078
ISBN: 9781949109559

Printed in the United States

Permission granted to copy activity pages at the end of the book.

Dedication

This book was written for Luca, Addison, and Jemma; Auggie, Alayna, and Lacey, my great grandchildren. It was written with the hope that this would start them on a path of love - loving God and loving reading.

My name is Luca. This is my sister Addy.

Addy loves unicorns, and I love dinosaurs.

Unicorns are pretend, but God really made dinosaurs.

I can almost hear them roar!

God made us, and God made you.

God made everything, the dinosaurs, too.

I know that's true.

Come with Addy & me to see a few.

No one has ever seen a live dinosaur.

God made them so long ago.

Dinosaur bones are found in the earth below.

Paleontologists dig up fossils, that's how we know.

Definitions

fossils: remains (bones and things) of animals and plants that lived long ago

paleontologist: person who studies ancient life, like dinosaurs

Carnivores ate meat. Herbivores ate plants in the glade.

But omnivores ate meat **and** plants in the shade.

Come with Addy and me; don't be afraid,

To see the amazing dinosaurs God made.

Definitions

carnivore: animal that eats meat

herbivore: animal that eats plants

omnivore: animal that eats meat and plants

Did you know?

Some dinosaurs were fast, and some were slow.

Some walked on 2 legs, and some walked on 4 legs.

But 2 legs or 4, fast or slow, they all came from eggs.

How Big Were Dinosaurs?
Small: up to 15 feet
Big: 15 to 30 feet
Huge: 30 to 65 feet
Gigantic: over 65

The *Brontosaurus* was not fierce when he played.

But he was big. Can you guess how much he weighed?

At 17 tons, he didn't need to be afraid.

His size caused enemies to be dismayed.

Look, at the dinosaur that is next.

The king of dinosaurs, *Tyrannosaurus* rex.

He was fierce and would fight with all his might.

Others stayed away because he would bite!

The *Ankylosaurus* was big but had tiny teeth.

So, he did not eat meat.

He was slow, but armor protected him from any foe.

And his clubbed tail could give a deadly blow.

I love my *Triceratops*,

Look at his head with three horns on top.

The horns helped him to fight.

And gave smaller dinosaurs a fright.

Yeah, for the *Stegosaurus*,

He was bigger than my school bus.

Look at all the tall plates on his back.

He used his spikey tail to give enemies a whack.

The *Carnotaurus* looked scary.

If I saw him, I would not tarry.

'Cause those sharp teeth look dangerous.

He was not a dinosaur you could trust.

The *Velociraptor* used his claws to catch prey.

He was a predator who would strike without delay.

This turkey-sized dinosaur had feathers,

But he couldn't fly, even in good weather.

Pterodactyls loved to fly in style.

Not really dinosaurs, they were scary, flying reptiles.

Little fish swim away, don't look back.

The pterodactyls are looking for a snack.

The *Parasaurolophus* looks kind of funny.

Even though he weighs 2 tons, he's cute as a bunny.

The crest on his head,

Was for making noise, it is said.

The *Harpymimus* was an omnivore,

Eating meat and plants galore.

Looking for food, he would explore,

In the forest and near the shore.

The *Brachiosaurus* ate lots of leaves.

Forty-two feet tall, he could eat from the tops of trees.

He weighed 120,000 pounds.

So, he walked with all 4 feet on the ground.

The *Diplodocus* had the longest tail of all.

No one would ever say he was small.

His front legs were shorter, but he didn't fall.

And he didn't eat from treetops, even though he was tall.

The *Oviraptor* was an omnivore.

He had feathers and walked on two feet not four.

He had no teeth but a sharp beak,

That helped him eat plants or meat.

I love the *Pachycephalosaurus.*

Call him Pachy – and don't make a fuss.

His hard-head was good protection, as you can see.

So, from his enemies he did not flee.

The *Giganotosaurus* is huge, I know.

It was the biggest carnivore, running to and fro.

Little creatures, your faces you should not show.

You will be a snack, if you are slow.

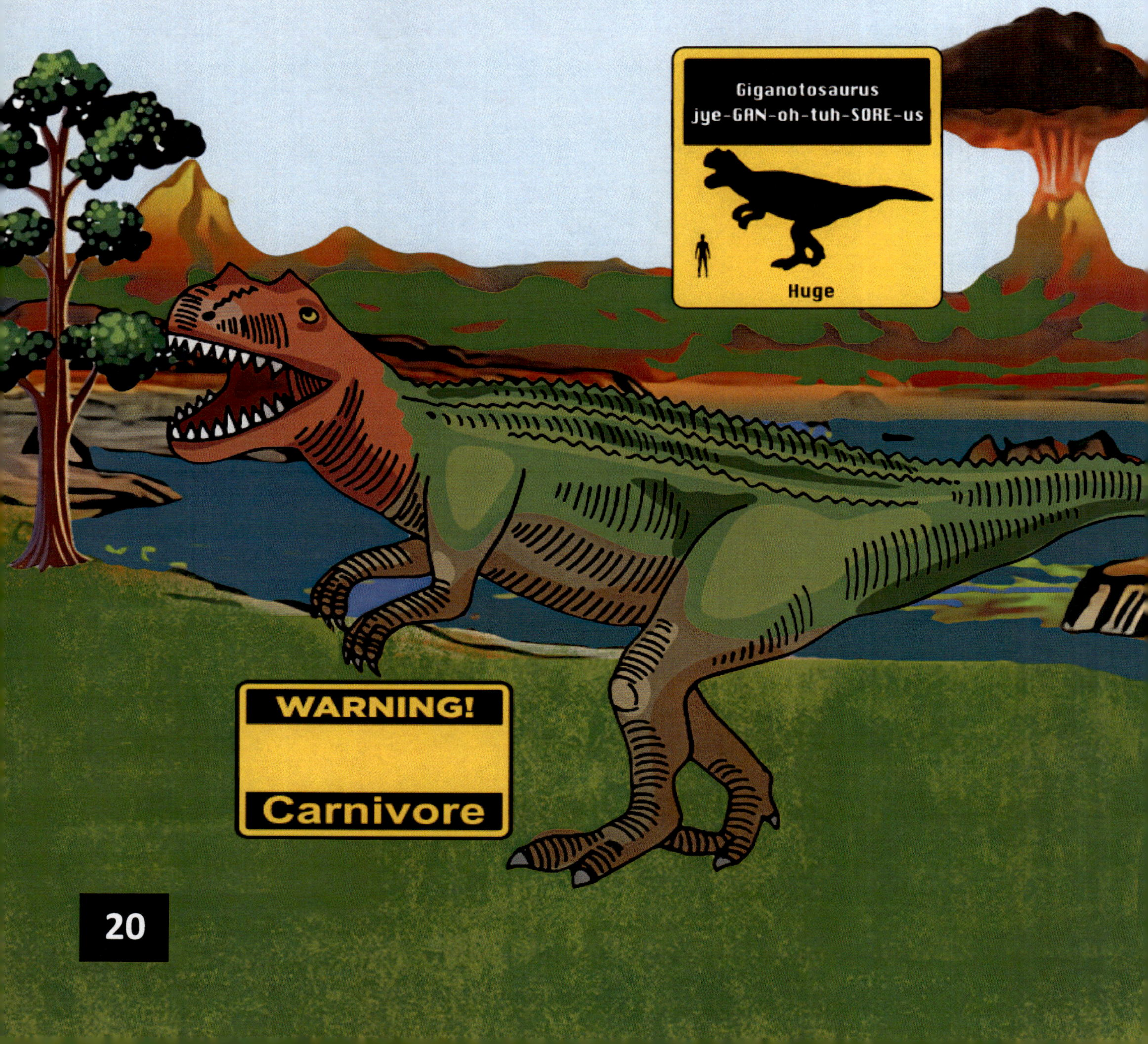

The *Spinosaurus*

Was dangerous.

Little creatures didn't want to be near his side.

So, they ran away to hide.

An *Iguanodon* sounds scary to me.

He was so big; I would have come up to his knee.

But he ate from little bushes and trees.

He wouldn't look so scary munching on leaves!

Dinosaurs are extinct.

We don't really know what to think.

Some think there were great dinosaur wars.

I don't know but there are no more dinosaurs.

I know God was creative, that's for sure,
When He made so many dinosaurs.
Even though they lived long ago,
God left the bones because, He wanted us to know.

Glossary

carnivore: an animal that eats meat

claw: sharp nail on hand or foot

dinosaur: group of extinct animals from long ago

extinct: no longer living

fierce: wild, savage, and mean

flee: to run away from danger

foe: enemy

fossils: remains of plants and animals that lived long ago

glade: open space in the forest

herbivore: an animal that eats only plants

omnivore: an animal that eats plants and meat

paleontologist: a person who studies ancient life

predator: an animal that hunts other animals for food

prey: an animal that is hunted and eaten by other animals

tarry: to stay somewhere

Dinosaur Sizes

Small: up to 15 feet:
Compsognathus, Deinonychus, Einiosuaurus, Falcarius, Guanlong, Harpymimus, Juravenator, Kentrosaurus, Leptoceratops, Oviraptor, Unaeesaurus, Velociraptor, Xianglong, Yinlong, Zalmoxes

Velocirator

Big: 15 to 30 feet:
Baryonyx, Carnotosaurus, Gallimimus, Icthyovenator, Nodosaurus, Pachycephalosaurus, Regaliceratops, Stegosaurus, Styracosaurus, Triceratops, Vulcanodon, Wuerhosaurus, Xenoceratops

Triceratops

Huge: 30 to 65 feet:

Allosaurus, Ankylosaurus, Iguanodon, Maiasuara, Olorotitan, Parasaurolophus, Quetza'coatlus, Spinosaurus, Tryannosaurus

Spinosaurus

Gigantic: over 65 feet

Apatosaurus, Barosaurus, Brachiosaurus, Brontosaurus, Diplodocus, Drednoughts, Giganotosaurus

Brontosaurus

Dinosaurs from A to Z

A

Apatosaurus
uh-paa-tuh-SORE-us

B

Barosaurus
bar-oh-SORE-us

C

Cerotosaurus
cer-ah-tow-SORE-us

D

Drednoughts
DRED-nawts

E

Einiosaurus
EYE-nee-oh-SORE-us

F

Falcarius
FAL-ca-REE-is

G

Guanlong
GWON-long

H

Harpymimus
har-pee-MIEM-us

I

Ichthyovenator
ICK-thee-oh-VEN-uh-tore

J

Juravenator
ju-rah-ve-nay-tore

K

Kentrosaurus
ken-trow-SORE-us

L

Leptoceratops
lep-toh-ser-ah-tops

M

Maiasaura

MY-uh-SORE-a

N

Nodosaurus

NOD-oh-SORE-us

O

Olorotitan

oh-lo-ro-TIE-ten

P

Parasaurolophus

PARA-sore-oh-la-fuss

Q

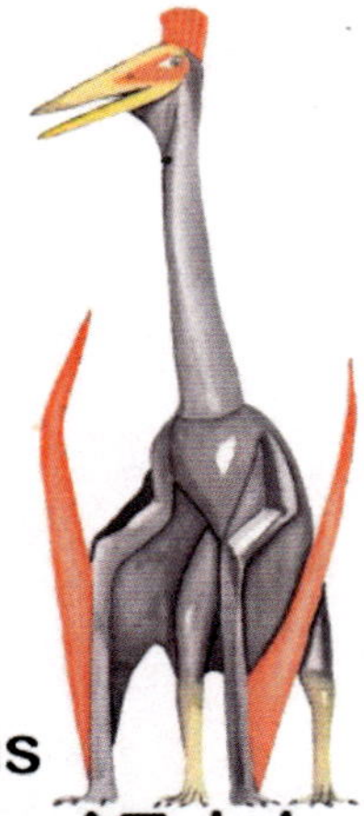

Quetzalcoatlus

ket-suhl-kow-AT-luhs

R

Regaliceratops

REE-ga-li-sera-tops

S

Stegosaurus

STEG-oh-SORE-us

T

Tyrannosaurus

tye-RAN-us-SORE-us

U

Unaysaurus

uh-na-ee-SORE-us

V

Vulcanodon

vul-kan-oh-don

W

Wuerhosaurus

where-oh-SORE-us

X

Xenoceratops

ZEN-oh-sera-tops

Y	Z

Yinlong	Zalmoxes
yin-long	zal-mox-is

- Copy cards on Cardstock.
- Cut out individual cards.

Things to do with cards:

Google to learn more about each dinosaur.

Make 2 sets and play card games like *go fish* or memory games.

Put in alphabetical order (for young children only use 3 to 5 cards at a time).

Some dinosaurs are in the book. Look for them.

Dinosaurs to Color

TRICERATOPS

TYRANNOSAURUS REX

ALLOSAURUS

PACHYCEPHALOSAURUS

BARYONYX

CERATOSAURUS

PARASAUROLOPHUS

COMPSOGNATHUS

SPINOSAURUS

OVIRAPTOR

IGUANODON

DEINONYCHUS

GALLIMIMUS

STYRACOSAURUS

STEGOSAURUS

How To Draw A Dinosaur

YOUR TURN

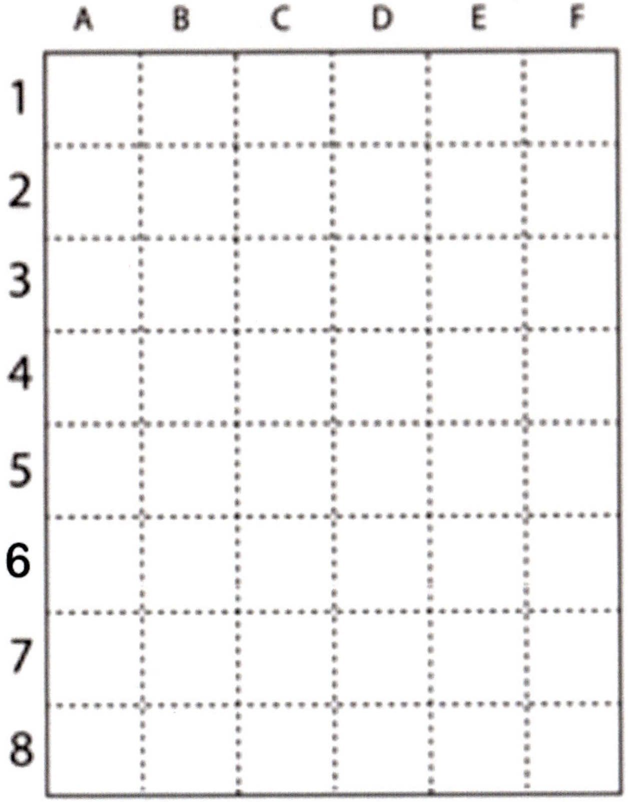

How To Draw A Dinosaur

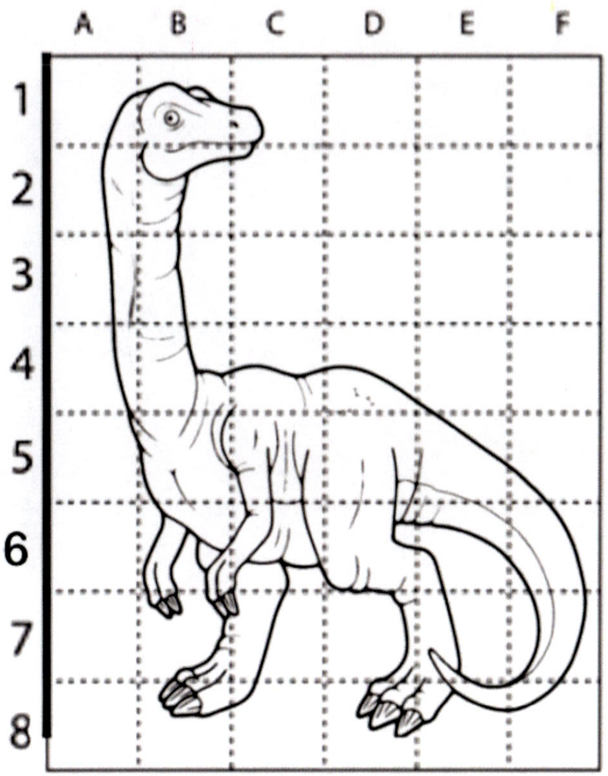

YOUR TURN

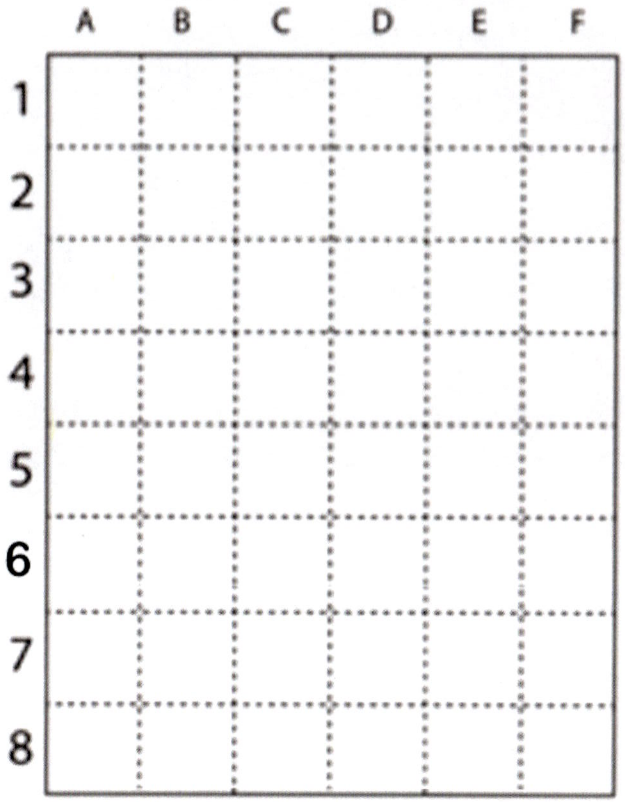

Pronunciation and Fact Sheet

Dinosaur	Pronunciation	Size	Length	Diet
Allosaurus	al-loh-SORE-us	H	39 feet	C
Ankylosaurus	AN-kee-low-SORE-us	H	34 feet	H
Apatosaurus	ah-PAT-oh-SORE-us	G	69 feet	H
Barosaurus	BAR-oh-SORE-us	G	89 feet	H
Baryonyk	bar-ee-ON-ICKS	B	31 feet	C
Brachiosaurus	BRACK-ee-oh-SORE-us	H	82 feet	H
Brontosaurus	BRONT-uh-SORE-us	H	72 feet	H
Carnotaurus	KAR-no-TORE-us	B	25 feet	C
Ceratosaurus	SER-a-toe-SORE-us	B	20 feet	C
Compsognathus	KOMP-sog-NAH-thus	S	4 feet	C
Deinonychus	DIE-no-NIKE-us	S	10 feet	C
Diplodocus	dih-PLOD-oh-kus	G	89 feet	H
Dreadnoughts	DRED-nawts	G	85 feet	H
Einiosaurus	EYE-nee-oh-SORE-us	B	20 feet	H
Falcaruis	FAL-ca-REE-us	S	12 feet	H
Gallimimus	GAL-ih-MIME-us	B	20 feet	O
Giganotosaurus	gi-GAN-oh-toe-SORE-us	H	43 feet	C
Guanlong	GWON-long	S	11 feet	C
Harpymimus	Har-pee-MIME-us	S	7 feet	O
Ichthyovenator	ICK-thee-oh-VEN-ah-tore	B	28 feet	C
Iguanodon	ig-WA-no-DON	H	40 feet	H
Juravenator	joor-oh-ven-ah-tore	S	2 feet	C
Kentrosaurus	ken-tro-SORE-us	S	15 feet	H
Leptoceratops	lep-toh-cer-ah-tops	S	6 feet	H
Maiasaura	MY-ah-SORE-ah	H	30 feet	H
Nodosaurus	NOD-o-SORE-us	B	20 feet	H
Olorotitan	oh-lo-ro-TIE-ten	H	33 feet	H

Dinosaur	**Pronunciation**	**Size**	**Lenth**	**Diet**
Oviraptor	OH-vih-RAP-tore	S	3 feet	O
Pachycephalosaurs	pack-ee-SEF-uh-lo-SORE-us	B	16 feet	H
Parasaurolophus	PARA-sore-oh-la-fuss	H	40 feet	H
Pteranodon (reptile)	ter-AN-uh-DON	S	20 ft.	C
Pterodactyl (reptile)	ter-uh-DAK-tuhl	S	3 feet	C
Quetzalcoatlus(reptile)	ket-zal-ko-AT-lus	H	22 feet	C
Regaliceratops	REE-ga-li-sera-tops	B	16 feet	H
Spinosaurus	SPINE-oh-SORE-us	H	50 feet	C
Stegosaurus	STEG-oh-SORE-us	B	30 feet	H
Styracosaurus	sty-RACK-oh-SORE-us	B	18 feet	H
Triceratops	try-SERR-uh-tops	B	30 feet	H
Tyrannosaurus	tie-RAN-oh-SORE-us	H	40 feet	C
Unaysaurus	oo-nay-SORE-us	S	8 feet	H
Velociraptor	vel-uh-si-RAP-tore	S	6 feet	C
Vulcanodon	vul-KAN-oh-don	B	21 feet	H
Wuerhosaurus	where-ho-SORE-us	B	27 feet	H
Xenoceratops	ZEN-oh-sera-tops	B	20 feet	H
Yinlong	yin-long	S	3 feet	H
Zalmoxes	zal-mox-is	S	10 inch	H

Size: Small up to 15 feet; Big 15 – 30 feet; Huge 30 to 65 feet; Gigantic over 65 feet

Diet: C=carnivore H=herbivore O=omnivore

Flying Reptiles: Size is wingspan

Index

About the Author

Carol Pirog has been an educator for many years. Working with young children and helping them to learn has been a life-time goal. She started writing the **This I Know** Series of books for her great grandchildren. She lives in the Midwest with her husband, John, and her dog, Bailey. Carol and her husband look forward to opportunities to visit their great grandchildren.

Other Books by Carol Pirog

God Made the Animals at the Zoo is the first book in the **This I Know** Series. **This I Know** is a series of books to help parents teach children that God is the Creator. The basic text of the book is written for young children or beginning readers. Inset boxes allow the book to 'grow' with the child. Boxes include text that is more advanced, and concepts that are more complex. While reading parents are following the admonition to incorporate teaching about God in daily activities. Whether reading the book or engaging in suggested activities, God can always be part of the activity.

God Made the Dogs I Love: A Book of Opposites is Book 2 in the **This I Know** Series. The book helps parents teach children that God is the Creator while teaching opposites in a fun way. Following the admonition to incorporate teaching about God in daily activities with your children, whether you are reading a book or learning some other concept, God can always be part of the activity.

No Sweat Alphabet for young children is a book about making learning fun. Research shows that children learn best when they are engaged. The book has fun games and activities to help children engage in the task of learning the alphabet. The idea of learning being fun will start your child on a lifelong path of learning.

Made in United States
North Haven, CT
26 February 2026

89130509R00027